Ocean Life Up Close

Walruses

by Kari Schuetz

MW00593124

BLASTOFF!
3
READERS

BELLWETHER MEDIA • MINNEAPOLIS, MN

BLASTOFF!
READERS

Note to Librarians, Teachers, and Parents:

Blastoff! Readers are carefully developed by literacy experts and combine standards-based content with developmentally appropriate text.

Level 1 provides the most support through repetition of high-frequency words, light text, predictable sentence patterns, and strong visual support.

Level 2 offers early readers a bit more challenge through varied simple sentences, increased text load, and less repetition of high-frequency words.

Level 3 advances early-fluent readers toward fluency through increased text and concept load, less reliance on visuals, longer sentences, and more literary language.

Level 4 builds reading stamina by providing more text per page, increased use of punctuation, greater variation in sentence patterns, and increasingly challenging vocabulary.

Level 5 encourages children to move from "learning to read" to "reading to learn" by providing even more text, varied writing styles, and less familiar topics.

Whichever book is right for your reader, Blastoff! Readers are the perfect books to build confidence and encourage a love of reading that will last a lifetime!

This edition first published in 2017 by Bellwether Media, Inc.

No part of this publication may be reproduced in whole or in part without written permission of the publisher. For information regarding permission, write to Bellwether Media, Inc., Attention: Permissions Department, 6012 Blue Circle Drive, Minnetonka, MN 55343.

Library of Congress Cataloging-in-Publication Data

Names: Schuetz, Kari, author.
Title: Walruses / by Kari Schuetz.
Description: Minneapolis, MN : Bellwether Media, Inc., [2017] | Series: Blastoff! Readers. Ocean Life Up Close | Audience: Age 5-8. | Audience: Grade K to grade 3. | Includes bibliographical references and index.
Identifiers: LCCN 2015051072 | ISBN 9781626174245 (hardcover : alk. paper) | ISBN 9781648346910 (paperback : alk. paper)
Subjects: LCSH: Walrus–Juvenile literature.
Classification: LCC QL737.P62 S38 2017 | DDC 599.79/9–dc23
LC record available at http://lccn.loc.gov/2015051072

Text copyright © 2017 by Bellwether Media, Inc. BLASTOFF! READERS and associated logos are trademarks and/or registered trademarks of Bellwether Media, Inc.

Printed in the United States of America, North Mankato, MN.

Table of Contents

What Are Walruses?

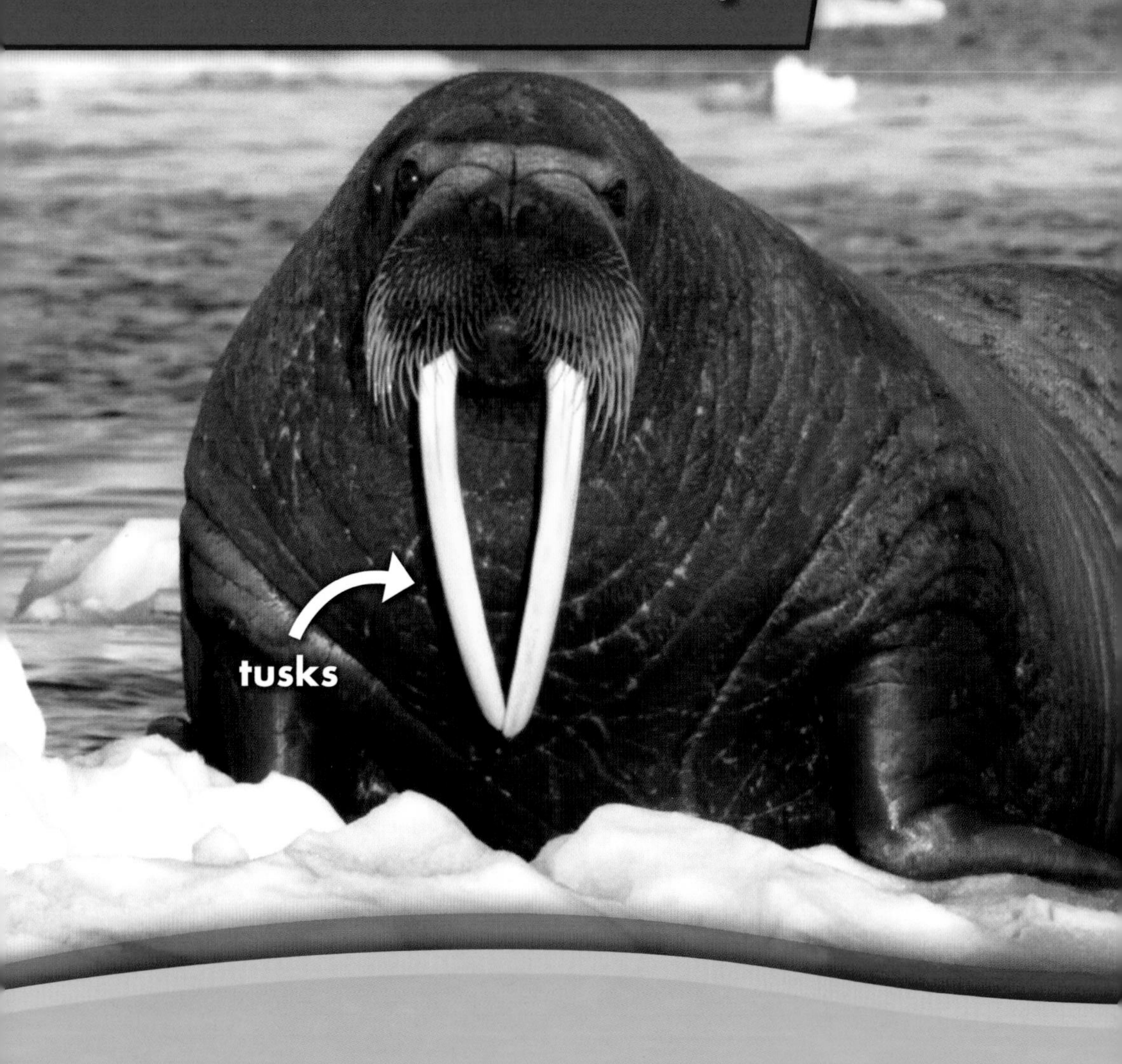

tusks

Walruses are **mammals** with mustaches. Two huge teeth called **tusks** stick out of their mouths.

Other Pinnipeds
earless seals
sea lions
fur seals

Like other **pinnipeds**, walruses have **flippers** for feet. Flippers help them move underwater and on ice.

These mammals are found in the **Arctic**. They mainly live in the cold, northern waters of the Pacific and Atlantic Oceans.

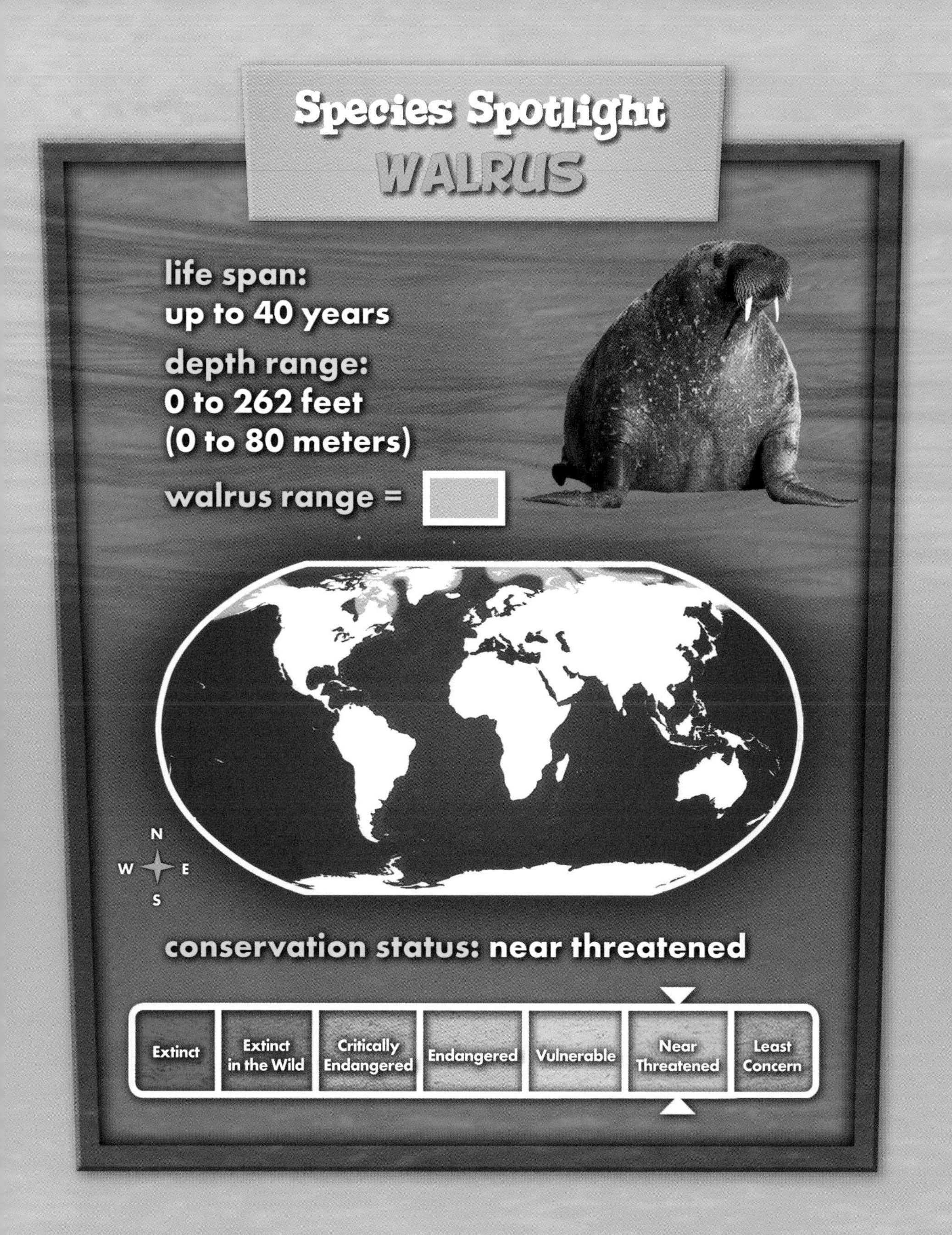
Species Spotlight
WALRUS
life span:
up to 40 years
depth range:
0 to 262 feet
(0 to 80 meters)
walrus range =
N
W
E
S
conservation status: near threatened
Extinct
Extinct in the Wild
Critically Endangered
Endangered
Vulnerable
Near Threatened
Least Concern

The large animals rest on land and floating ice.

Blubber and Whiskers

Walruses are huge in size. The biggest males can be 12 feet (3.6 meters) long and more than 3,000 pounds (1,361 kilograms)!

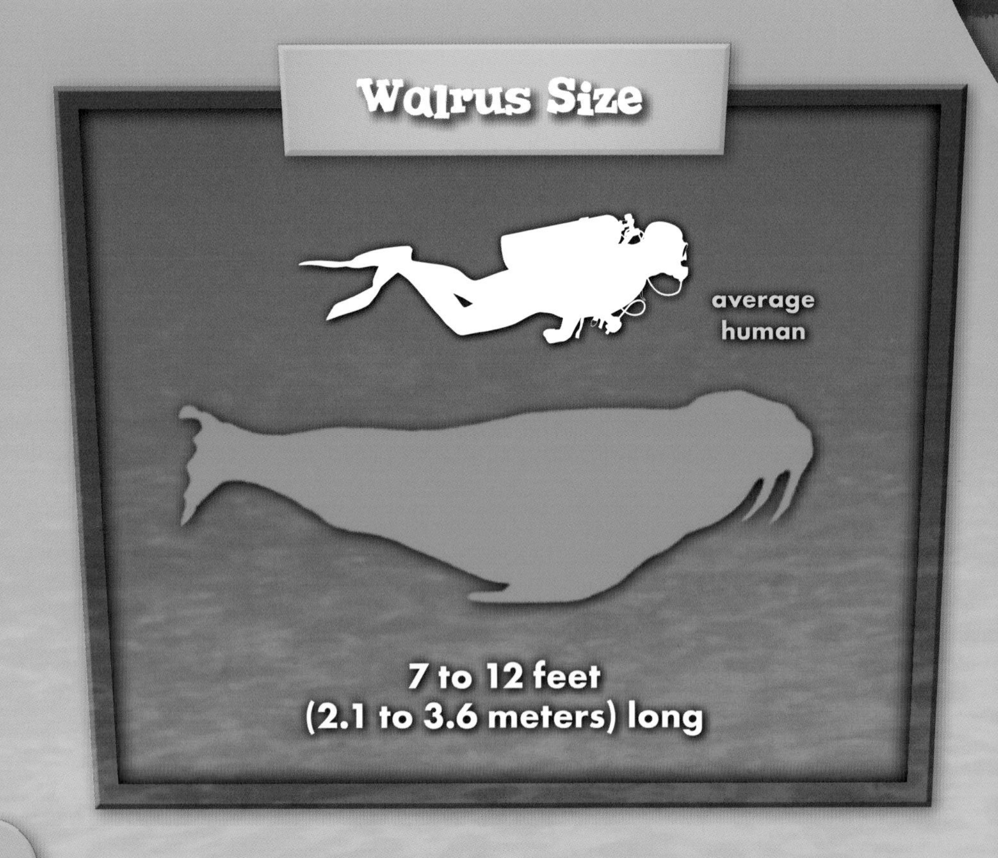
Walrus Size
average human
7 to 12 feet
(2.1 to 3.6 meters) long

Part of their weight is thick **blubber**. This layer of fat under their skin keeps them warm.

Walruses' tough skin is wrinkled. It is usually gray or brown in the cold water. But it can turn pinkish on land.

Walruses usually suck up **prey** from the ocean floor. Whiskers help them feel for food.

Catch of the Day

soft-shell
clams

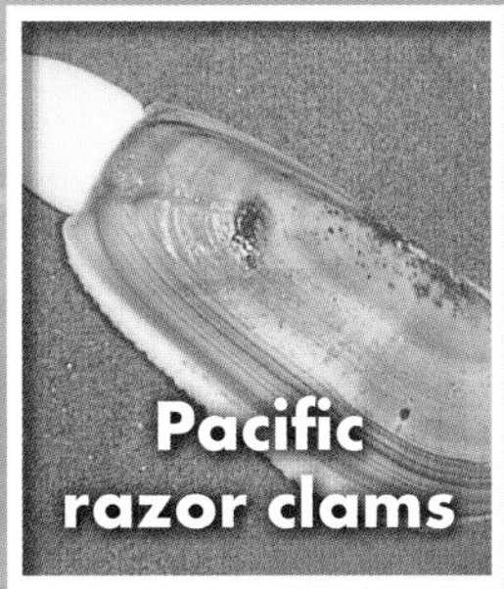
Pacific
razor clams

Baltic
clams

whiskers

Front flippers work like paddles in water. They guide turns. Back flippers power walruses forward.

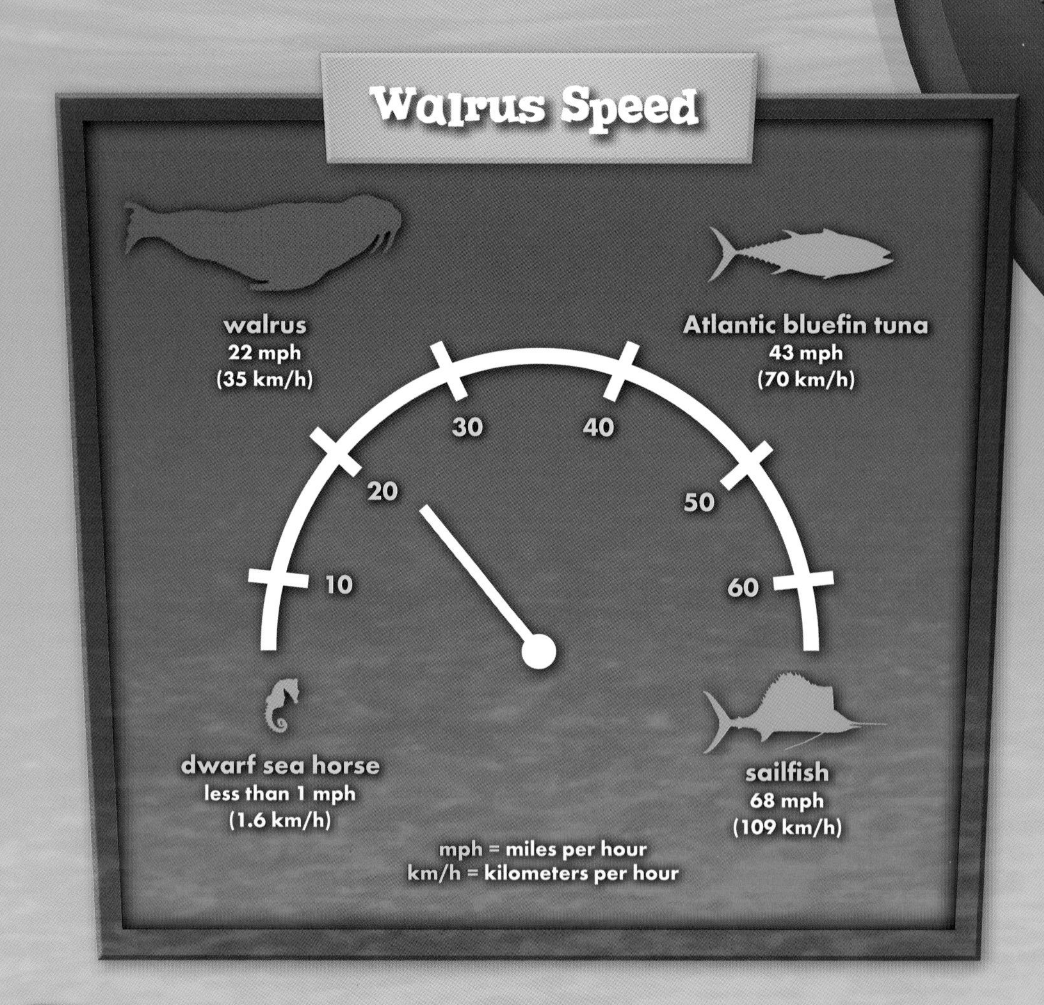
Walrus Speed
walrus
22 mph
(35 km/h)
Atlantic bluefin tuna
43 mph
(70 km/h)
10
20
30
40
50
60
dwarf sea horse
less than 1 mph
(1.6 km/h)
sailfish
68 mph
(109 km/h)
mph = miles per hour
km/h = kilometers per hour

front
flipper
back
flipper

Walruses also use flippers to flop on land. Rough bottoms keep the flippers from slipping on ice.

Tough Tusks

Up to 3 feet (1 meter) long, walrus tusks are strong tools. They can chip at ice to make holes for breathing.

Identify a Walrus
thick mustache
long tusks
flat flippers

Walruses also use tusks like ice picks. These teeth help lift walruses out of the water!

Males have longer tusks than females. They show them off in **threat displays**.

Sometimes the tusks become weapons to fight other males. The animals fight over **territory** and females.

In the Herd

Hundreds of walruses can make up a **herd**. Male and female walruses spend most of the year in separate herds.

herd

Pups grow up with females. The moms look after the pups for up to three years.

Male walruses always make sure they are heard. They often growl when they fight.

To charm females, males make bell-like sounds underwater. To do this, pouches on their necks swell with air!

Glossary

Arctic—the cold region around the North Pole

blubber—the fat of walruses

flippers—flat, wide body parts that are used for swimming

herd—a group of walruses

mammals—warm-blooded animals that have backbones and feed their young milk

pinnipeds—ocean mammals with four flippers; sea lions, seals, and walruses are pinnipeds.

prey—animals that are hunted by other animals for food

pups—baby walruses

territory—the area where an animal lives

threat displays—behaviors that animals perform to show strength

tusks—long, curved teeth

To Learn More

AT THE LIBRARY

Kaufmann, Carol. *Polar: A Photicular Book*. New York, N.Y.: Workman Publishing, 2015.

Miller, Sara Swan. *Walruses of the Arctic*. New York, N.Y.: PowerKids Press, 2009.

Person, Stephen. *Walrus: Tusk, Tusk*. New York, N.Y.: Bearport Pub., 2011.

ON THE WEB

Learning more about walruses is as easy as 1, 2, 3.

WWW.FACTSURFER.COM

1. Go to www.factsurfer.com.
2. Enter "walruses" into the search box.
3. Click the "Surf" button and you will see a list of related web sites.

With factsurfer.com, finding more information is just a click away.

Index

The images in this book are reproduced through the courtesy of: Vladimir Melnik, front cover, pp. 3, 7, 15 (bottom); Louise Murray/ Age Fotostock/ SuperStock, pp. 4-5; Marco Rolleman, p. 5 (top); Longjourneys, p. 5 (center); SkyLynx, p. 5 (bottom); SuperStock/ Glow Images, p. 6; Michael Nolan/ robertharding/ SuperStock, p. 9; Yuriy Kvach/ Wikipedia, p. 11 (top left); RazorClam23/ Wikipedia, p. 11 (top center); Sandy Rae/ Wikipedia, p. 11 (top right); Rebecca Jackrel/ Age Fotostock/ SuperStock, p. 11 (bottom); Paul Souders/ Corbis, pp. 13, 16; Kenneth Canning, p. 14; SasinT, p. 15 (top left); tryton2011, p. 15 (top center); Hal Brindley, p. 15 (top right); Juniors Bildarchiv GmbH/ Alamy, p. 17; Maximilian Buzun, p. 18; tryton2011, p. 19; BMJ, p. 20; Fabrice Simon/ Corbis, p. 21.